THE ULTIMATE COMPETITIVE WISDOM

Mastering the Art of Winning and Thriving in Today's Dynamic World

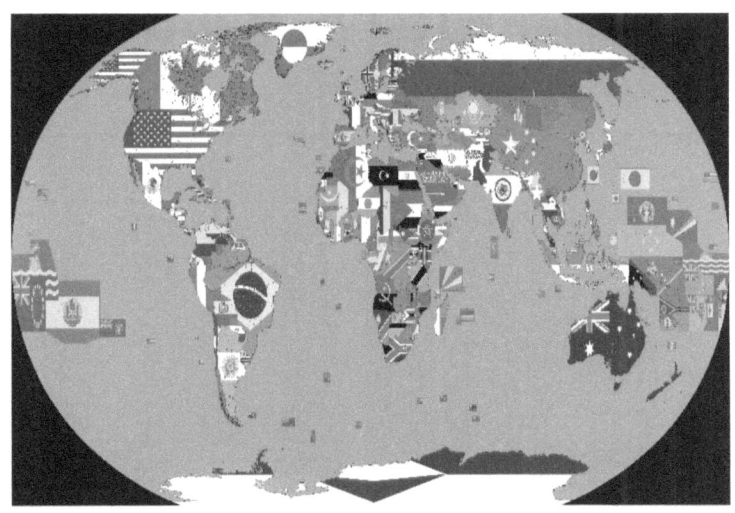

Henry M. Yim
Alexandra V. Yim

Authors of *WiseBook for Charismatic Leaders*, *Milky Way Poems*, and *The Great Game XXI*

**THE ULTIMATE
COMPETITIVE WISDOM**

THE ULTIMATE COMPETITIVE WISDOM

Copyright © 2022 Henry M. Yim and
Alexandra V. Yim
All rights reserved.

DEDICATION

To all the competitive strategists and global leaders of corporations, governments, non-profits, and individuals who work tirelessly to transform innovation and create a sustainable and prosperous future for all, this book is dedicated to you.

To Dad, who instilled in us the importance of creative thinking, a transformative mindset, enthusiasm for hard work, and continuous learning, and to Mom, who inspired us to write, maintain integrity, and serve others in the community, thank you both for your unwavering support and for being our pillars of strength throughout this journey.

To Wesley, your incredible work as a medical doctor during the challenging, historic pandemic is deeply appreciated. Your unwavering dedication to saving lives in emergency medicine is remarkable and serves as a testament to your character. Your commitment to serving others, belief in medicine, and compassion for patients will continue to inspire us for years to come. The bravery and sacrifices you made in the face of a challenging and unprecedented situation will be remembered and serve as a lasting inspiration.

CONTENTS

	List of the Maxims of Competitive Wisdom	vi
	Acknowledgments	ix
	Introduction	x
1	Embracing Competitive Wisdom	1
2	Peace Through Diplomacy	5
3	Adaptability and Agility	7
4	Self-Awareness and Transformation	9
5	Alliance with Empathy	12
6	Ethical Strategic Deception	14
7	Strategic Unpredictability	17
8	Loyal Workforce Collaboration	20
9	Chaos Conceals Opportunity	22
10	Harmonizing Offense and Defense	24
11	Power of Design Thinking Innovation	27
12	Greatest Competitive Edge	29
13	Preparation and Decisiveness	31
14	Outmaneuvering Your Competition	34
15	Strategic Reconnaissance	37
16	Complacency Erodes Excellence	41
17	Synergy of Strategy and Tactics	44
18	Sacrifice and Success	48
19	Balancing Aggression and Protection	51
20	Art of Minimal Force	54
21	Achieving Exponential Success	57
	Epilogue: Pursuing the Infinite Quest	60
	References	62
	Index	66
	About the Authors	70

THE ULTIMATE COMPETITIVE WISDOM

LIST OF THE MAXIMS OF COMPETITIVE WISDOM

The Maxims of Competitive Wisdom

Maxim I: True victory in any conflict is to attain peace through strategic diplomacy and wise decision-making, avoiding the need for actual warfare. — 5

Maxim II: Success in business competition requires strategic adaptability. Anticipate your rival's moves, provoke their weaknesses, and exploit their blind spots. To win, stay agile and strike where they least expect it. — 7

Maxim III: Victory belongs to those who understand both their strengths and weaknesses as well as their opponents'. — 9

Maxim IV: Embrace your allies, study your rivals, and conquer the competition with strategic prowess — 12

Maxim V: Effective competition requires strategic deception. To win, project weakness when strong, appear idle when active, feign distance when close, and simulate proximity when distant. — 14

Maxim VI: Gain a competitive edge through strategic unpredictability. Diverge from their — 17 strengths, erode their confidence, leverage their weaknesses, and triumph with unforeseen approaches.

Maxim VII: Treat your team like family, and they'll brave any challenge with you. See them as cherished kin, and they'll remain fiercely loyal through thick and thin. — 20

Maxim VIII: Chaos conceals opportunity, for those with the wisdom to find it. — 22

Maxim IX: The path to invincibility is through a strong defense, while victory is achieved through a strategic offense. — 24

Maxim X: Limitations in resources do not hinder creativity or innovation. Strategic combinations of available resources can produce endless opportunities and possibilities, just like musical notes form new melodies. — 27

Maxim XI: Realizing our full potential unleashes our greatest competitive edge. — 29

Maxim XII: Victory is achieved before the competition begins. Success belongs to those who prepare meticulously and act decisively, while failure awaits those who enter the arena unprepared. — 31

THE ULTIMATE COMPETITIVE WISDOM

Maxim XIII: Winning in competition isn't about direct confrontation, but rather weakening your opponent's resolve through effective strategy and execution.   34

Maxim XIV: The visionary leader and astute strategist harness the collective intelligence of their team, employing it for strategic reconnaissance, ultimately achieving extraordinary outcomes.   37

Maxim XV: Complacency erodes excellence, leaving greatness to wither.   41

Maxim XVI: The synergy of strategy and tactics is the cornerstone of triumph; one without the other leads to failure.   44

Maxim XVII: Success favors those who embrace sacrifice.   48

Maxim XVIII: Unleash offense from celestial heights, shield defense in subterranean depths.   51

Maxim XIX: Unleash the power of the few, conquer the impossible.   54

Maxim XX: Seize opportunities, multiply success.   57

ACKNOWLEDGMENTS

We express our gratitude to our teachers, friends, and the numerous international scholars whose invaluable insights, feedback, and assistance have played a crucial role in the research and writing of this book. Your contributions have been pivotal in shaping our ideas and fortifying our perspective of the world.

THE ULTIMATE COMPETITIVE WISDOM

INTRODUCTION

In today's fast-paced and highly competitive world, success goes beyond mere strategy or the latest technology. It hinges on a profound understanding of timeless principles of competition and their contextual application.

"The Ultimate Competitive Wisdom" stands as a testament to our extensive research on competitive stratagems, meticulously synthesized to embody the essence of competitive wisdom in this amalgamation of maxims. Drawing inspiration from Sun Tzu's renowned masterpiece, "The Art of War," we have delved into a wealth of knowledge and expertise, ingeniously fusing Sun Tzu's profound teachings with contemporary principles of transformational leadership and design thinking. This adaptation allows us to incorporate and infuse these principles effectively in the complexities of the modern world, transcending boundaries and offering invaluable insights applicable to diverse domains such as business, government, and everyday life. In this context, "The Ultimate Competitive Wisdom" vividly illustrates the paramount importance of competitive wisdom in today's dynamic and fast-paced world, drawing inspiration from Sun Tzu's teachings while highlighting its practical applicability and relevance in modern-day scenarios. Through its synthesis of traditional wisdom and contemporary insights, this comprehensive guide goes beyond a mere rehashing of Sun Tzu's work, providing a unique perspective on competitive strategies and emphasizing their significance in today's competitive landscape.

THE ULTIMATE COMPETITIVE WISDOM

Through insightful analysis and real-life scenarios, this book takes readers on a journey through the fundamental maxims of competition. It emphasizes the importance of strategic thinking, the power of agility and adaptability, and the value of collaboration and cooperation. By deeply understanding and leveraging these maxims, readers can gain a competitive edge in the realms of business, government, and daily life.

"The Ultimate Competitive Wisdom" is not solely intended for business executives or military leaders; it caters to anyone striving for success and seeking a competitive advantage in any field of endeavor. Whether you're an entrepreneur aiming to grow your business, a government official aspiring to implement effective policies, or simply an individual desiring personal excellence, this book equips you with the necessary tools and insights for success.

If you're ready to unleash your competitive edge and elevate your performance to new heights, "The Ultimate Competitive Wisdom" is the book for you. It's time to draw from time-proven wisdom and apply it to the challenges of the present and the future.

Henry M. Yim
Alexandra V. Yim
Washington, D.C., USA
July 4, 2022

THE ULTIMATE COMPETITIVE WISDOM

"To be prepared for war is one of the most effectual means of preserving peace." – George Washington

1
EMBRACING COMPETITIVE WISDOM

Competitive wisdom is the art of employing methods and techniques to gain an advantage over rivals. In today's fiercely competitive global landscape, developing a robust competitive strategy is essential for achieving success in any domain.

"The Art of War" by Sun Tzu, an esteemed Chinese strategist, is widely regarded as one of the most profound works on military strategy. Its relevance extends beyond the realm of warfare, as it offers valuable insights applicable to business, government, and everyday life. Drawing inspiration from Sun Tzu's teachings, as well as contemporary principles of transformational leadership and design thinking, "The Ultimate Competitive Wisdom" illustrates the significant importance of competitive wisdom in the context of modern-day scenarios.

Within the realm of business, possessing a competitive edge is pivotal to success. Companies that fail to keep pace with their competitors often falter and eventually fade away. By mastering the art of competitive wisdom, businesses can outshine their rivals, attract more customers, and enhance their profitability. The book showcases numerous examples of successful companies that have harnessed competitive strategies to achieve their goals.

Competitive strategy is equally crucial for governments. In today's world, nations compete for resources, talent, and investments. By implementing effective policies and strategies, governments can attract greater investments, generate more jobs, and bolster their economies. The book highlights instances of governments that have employed competitive strategies to foster economic growth and attain success.

Moreover, competitive wisdom can be applied in various facets of daily life. Whether it's securing a job, finding a partner, or achieving personal goals, having a competitive edge can make a significant difference. The book provides practical advice and tips

THE ULTIMATE COMPETITIVE WISDOM

on how to apply competitive wisdom to daily life situations.

Competitive wisdom embodies a range of principles and strategies that have withstood the test of time. These principles are universally applicable, extending their benefits to businesses, governments, and individuals alike. The following critical characteristics demonstrate the advantages of embracing time-proven competitive wisdom:

Strategy Development: Competitive wisdom offers a framework for crafting effective strategies. By comprehending the fundamental principles of competition, individuals, businesses, and governments can evaluate their strengths and weaknesses, identify opportunities and threats, and devise well-informed action plans.

Competitive Advantage: Competitive wisdom enables the identification and cultivation of unique strengths and advantages. By leveraging these distinctive qualities, businesses can differentiate themselves from competitors and gain a competitive edge. Governments can focus on areas where they possess comparative advantages to attract investments and bolster economic growth.

Adaptability and Resilience: Competitive wisdom emphasizes the importance of adaptability and resilience in the face of change. This mindset encourages individuals, businesses, and governments to continuously assess and adjust their strategies to stay ahead of the competition. The ability to adapt swiftly to new challenges and seize emerging opportunities paves the way for long-term success.

Resource Optimization: Effective competitive strategies prioritize the efficient allocation and utilization of resources. Businesses and governments can benefit from optimizing their resource allocation, be it financial capital, human resources, or technological capabilities. By making the most of available resources, organizations can achieve higher productivity and attain better outcomes.

Informed Decision-Making: Competitive wisdom underscores the

THE ULTIMATE COMPETITIVE WISDOM

significance of informed decision-making. By analyzing the competitive landscape, understanding market dynamics, and considering potential risks and rewards, individuals and organizations can make strategic and calculated choices. This approach minimizes the likelihood of hasty or ill-informed decisions, leading to superior outcomes.

Collaboration and Cooperation: Competitive wisdom recognizes the value of collaboration and cooperation. In our interconnected world, partnerships and alliances provide access to complementary strengths and resources. By forging strategic collaborations, businesses and governments can enhance their competitiveness and achieve mutually beneficial outcomes.

Continuous Learning: Competitive wisdom emphasizes the importance of continuous learning and improvement. It encourages individuals and organizations to constantly seek knowledge, analyze trends, and stay updated with the latest developments in their respective fields. This commitment to ongoing learning enables the refinement of strategies and the identification of new opportunities for growth and success.

Transformational Leadership: Embracing transformational leadership involves inspiring and motivating teams to achieve extraordinary results. It goes beyond transactional management and focuses on creating a vision, fostering innovation, and empowering individuals. Transformational leaders inspire their teams by setting clear goals and communicating a compelling vision of the future. They encourage creativity and risk-taking, promoting an environment of continuous improvement and learning. Through their charisma and passion, they motivate others to exceed their own expectations and achieve remarkable outcomes. By embracing transformational leadership as a part of competitive wisdom, organizations can cultivate a culture of high performance, collaboration, and innovation, driving them towards success in today's dynamic and competitive landscape.

Design Thinking: Design thinking brings a human-centered approach to problem-solving and innovation. It encourages

THE ULTIMATE COMPETITIVE WISDOM

organizations to empathize with their customers, understand their needs and challenges, and design solutions that truly address those needs. Design thinking promotes a creative and iterative process, involving brainstorming, prototyping, and testing ideas to arrive at innovative solutions. It emphasizes collaboration and cross-functional teamwork, leveraging diverse perspectives and expertise. By incorporating design thinking into their approach, organizations can unlock new opportunities, create user-centric products and services, and differentiate themselves in the market. It encourages a mindset of continuous improvement and encourages organizations to challenge conventional thinking and explore new possibilities.

In summary, embracing competitive wisdom offers valuable guidance and principles that benefit businesses, governments, and individuals across various aspects of life. By applying these strategies, organizations and individuals can gain a competitive advantage, optimize resource allocation, make informed decisions, adapt to changing circumstances, foster collaboration, and continue to learn and grow. Furthermore, incorporating transformational leadership and design thinking as integral components of competitive wisdom allows organizations to cultivate a culture of innovation, collaboration, and continuous improvement. These approaches enable organizations to stay ahead in a rapidly changing environment, meet evolving customer needs, and drive sustainable growth and success. Integrating competitive wisdom, along with the principles of transformational leadership and design thinking, into one's approach contributes to long-term success and the achievement of goals in an ever-evolving world.

2
PEACE THROUGH DIPLOMACY

Maxim I: *True victory in any conflict is to attain peace through strategic diplomacy and wise decision-making, avoiding the need for actual warfare.*

Maxim of Competitive Wisdom

The thinking behind the wisdom that "true victory in any conflict is to attain peace through strategic diplomacy and wise decision-making, avoiding the need for actual warfare" is based on the belief that war should always be a last resort. It recognizes that conflicts can arise between nations, groups, or individuals, and that resolving these conflicts in a peaceful way is ultimately the best outcome for all parties involved.

By pursuing peace through diplomacy instead of resorting to war, the goal is to find a mutually acceptable solution that satisfies the needs and interests of both parties, without causing unnecessary harm, destruction, and loss of life. Diplomacy allows for discussions and negotiations to take place in a controlled and constructive environment, where each side can express their grievances, present their case, and work towards a compromise or agreement.

While diplomacy may take longer to reach a resolution than going to war, it can help prevent further escalation of violence and ultimately lead to a more stable and peaceful outcome. It also acknowledges that the consequences of war can be devastating and long-lasting, both for those involved and for innocent civilians caught in the crossfire.

Overall, the thinking behind this wisdom is that peaceful conflict resolution should always be pursued as the first and best option,

THE ULTIMATE COMPETITIVE WISDOM

with war only considered as a last resort when all other options have been exhausted.

Real-Life Scenario
The Cuban Missile Crisis

One real-life example that aligns with the thinking expressed by this wisdom would be the Cuban Missile Crisis in 1962. During this tense period of the Cold War, the United States and the Soviet Union were on the brink of a potential nuclear war over the placement of Soviet missiles in Cuba.

Rather than immediately resorting to military action, both sides engaged in intense diplomatic negotiations to find a peaceful resolution. Through back-channel communications and direct negotiations between U.S. President John F. Kennedy and Soviet Premier Nikita Khrushchev, a deal was eventually struck.

The resolution, known as the "Cuban Missile Crisis Agreement," involved a public commitment by the Soviet Union to dismantle its missile bases in Cuba, while the United States agreed not to invade Cuba and to remove its missiles from Turkey. The crisis was resolved peacefully, preventing a direct military confrontation and potentially catastrophic consequences.

This example demonstrates how the most desirable outcome, peace, was pursued through diplomatic means and negotiations, rather than resorting to actual war. It highlights the importance of strategic thinking, understanding the interests of both sides, and seeking a resolution that serves the long-term interests of all parties involved.

3
ADAPTABILITY AND AGILITY

Maxim II: *Success in business competition requires strategic adaptability. Anticipate your rival's moves, provoke their weaknesses, and exploit their blind spots. To win, stay agile and strike where they least expect it.*

Maxim of Competitive Wisdom

This wisdom suggests that success in business competition requires more than just having a good strategy. It emphasizes the importance of being adaptable and flexible in the face of changing circumstances and the moves of competitors.

To succeed in business competition, you must be able to anticipate your rival's moves and stay ahead of them. This means understanding your competition and their strengths and weaknesses, as well as being able to identify opportunities to provoke their weaknesses and exploit their blind spots.

To win, it's crucial to remain agile and be able to pivot quickly when necessary. This means being willing to change your strategy or approach if it's not working, and being able to strike where your competition least expects it. This requires a certain level of creativity and thinking outside the box to find innovative solutions to challenges and opportunities.

Overall, this wisdom emphasizes the importance of adaptability, creativity, and agility to succeed in business competition. By staying ahead of rivals and adapting quickly, you gain a market edge and achieve long-term success.

THE ULTIMATE COMPETITIVE WISDOM

Real-Life Scenario

Coca-Cola: A Model of Global Strategic Adaptability

A real-life example of a global business that applies this thinking is Coca-Cola, which has been in a long-standing battle with its main competitor, PepsiCo. Coca-Cola has demonstrated strategic adaptability by diversifying its product portfolio, expanding into emerging markets, and leveraging technology to enhance customer engagement.

For instance, Coca-Cola has introduced various beverage brands, such as Vitaminwater and Powerade, to appeal to health-conscious consumers and expand its market reach. The company has also focused on growing its presence in emerging markets like India, China, and Africa, where it faces tough competition from local players.

Moreover, Coca-Cola has invested heavily in digital marketing to engage with customers and build brand loyalty. The company has launched various mobile apps and social media campaigns to reach out to millennials and other tech-savvy consumers.

By anticipating its rival's moves, exploiting their blind spots, and staying agile, Coca-Cola has been able to maintain its market dominance and continue to grow in the highly competitive global beverage industry.

4
SELF-AWARENESS AND TRANSFORMATION

Maxim III: *Victory belongs to those who understand both their strengths and weaknesses as well as their opponents'.*

Maxim of Competitive Wisdom

This wisdom encapsulates the essence of competitive wisdom, highlighting the fundamental requirement for a comprehensive understanding of oneself and one's adversaries to achieve victory. It underscores the significance of self-awareness and strategic thinking in the pursuit of success.

By intimately knowing their own strengths and weaknesses, individuals can capitalize on their advantages while proactively addressing their vulnerabilities. This self-awareness enables them to make informed decisions, optimize their approach, and leverage their unique attributes to gain a competitive edge.

Simultaneously, understanding the strengths and weaknesses of opponents provides a strategic advantage. It allows individuals to anticipate their opponents' moves, identify potential vulnerabilities, and exploit opportunities. Armed with this knowledge, they can adopt defensive measures to protect against their opponents' strengths and proactively exploit their weaknesses.

In essence, this wisdom recognizes that victory in competitive endeavors extends beyond mere brute force or innate talent. It requires a sophisticated understanding of the intricate dynamics at play. Success emerges from a combination of self-awareness, strategic analysis, and the ability to adapt and respond effectively to the actions of opponents.

THE ULTIMATE COMPETITIVE WISDOM

Ultimately, those who embody this wisdom, gaining profound insights into themselves and their opponents, are poised for success in any competitive arena. By embracing this wisdom, individuals can navigate the complexities of competition with greater clarity and purpose, increasing their likelihood of achieving victory.

Real-Life Scenario
Samsung Vs. Apple

A compelling demonstration of this wisdom can be observed in the long-standing rivalry between Samsung and Apple in the smartphone industry. Both companies exhibit a deep understanding of their own strengths and weaknesses, as well as those of their competitors. Apple is widely recognized for its premium, high-end devices known for their sleek design, seamless user experience, and strong brand appeal. Meanwhile, Samsung has built a reputation for offering a wider range of devices across various price points, along with cutting-edge technology and a penchant for innovation.

Samsung, in particular, has shown remarkable adaptability and strategic acumen in response to the ever-evolving market dynamics. Recognizing the immense popularity of Apple's iPhone, Samsung strategically differentiated itself by introducing larger screen sizes and incorporating stylus pens in its Galaxy Note series. This move not only appealed to consumers seeking a different user experience but also established a unique selling point for Samsung devices. By understanding the demand for diversity and customization, Samsung expanded its product lineup to cater to different market segments, including budget-friendly options and devices with advanced features.

Furthermore, Samsung's marketing and advertising strategies have played a crucial role in challenging Apple's dominance. The company invested heavily in campaigns that directly compared its products to Apple's, highlighting features where Samsung excelled

THE ULTIMATE COMPETITIVE WISDOM

and emphasizing perceived weaknesses in Apple's offerings. This approach allowed Samsung to capture consumer attention and position itself as a viable alternative to Apple's devices.

Similarly, Apple has leveraged its strengths in design, user experience, and ecosystem integration to maintain its competitive edge. The company consistently introduces innovative features and technologies that captivate its loyal customer base. For instance, the introduction of the Face ID facial recognition system in the iPhone X showcased Apple's commitment to pushing boundaries and setting new industry standards. Apple's emphasis on creating a seamless ecosystem, integrating hardware, software, and services, has also played a significant role in fostering customer loyalty and differentiating its products from competitors.

By continuously assessing the market landscape, understanding customer preferences, and leveraging their respective strengths, both Samsung and Apple have managed to stay ahead in the fiercely competitive smartphone industry. Their ability to adapt, innovate, and respond to market trends has enabled them to dominate market share and cultivate dedicated customer bases worldwide.

In conclusion, the rivalry between Samsung and Apple serves as a prime example of how embracing competitive wisdom, including understanding strengths and weaknesses, strategic differentiation, and targeted marketing, can drive success in the dynamic world of smartphones. These companies' ability to adapt to changing market demands, introduce innovative features, and connect with consumers has solidified their positions as industry leaders.

5
ALLIANCE WITH EMPATHY

Maxim IV: *Embrace your allies, study your rivals, and conquer the competition with strategic prowess.*

Maxim of Competitive Wisdom

The competitive wisdom "Embrace your allies, study your rivals, and conquer the competition with strategic prowess" is a powerful strategy that can help individuals and organizations achieve competitive success.

The first component of the strategy, "Embrace your allies," highlights the importance of cultivating strong relationships with those who can support you in your competitive endeavors. In the business context, this might mean partnering with suppliers, distributors, or other companies in your industry. By building these relationships, you can gain access to valuable resources, knowledge, and networks that can help you compete more effectively.

The second component, "Study your rivals," emphasizes the importance of understanding your competition. By analyzing your competitors' strengths, weaknesses, and strategies, you can identify opportunities to differentiate yourself and gain a competitive advantage. This might involve researching their products or services, analyzing their marketing campaigns, or studying their financial performance.

The final component, "Conquer the competition with strategic prowess," brings the first two components together to create a powerful competitive strategy. By leveraging the strengths of your allies and capitalizing on the weaknesses of your rivals, you can develop a strategic plan that positions you for success. This might involve developing new products or services, entering new markets, or implementing innovative marketing strategies.

Overall, this competitive wisdom emphasizes the importance of building strong relationships, understanding your competition, and using strategic thinking to gain a competitive advantage. By embracing these principles, individuals and organizations can position themselves for long-term success in a competitive marketplace.

Real-Life Scenario
Strategic Alliance of Apple and IBM

One example of a business that has successfully applied this competitive wisdom is the collaboration between Apple and IBM. In 2014, Apple and IBM formed a strategic partnership that aimed to provide enterprise customers with mobile apps and services. Apple had the hardware and software expertise, while IBM had the enterprise-level solutions and services.

By collaborating with IBM, Apple was able to access the enterprise market, which it had previously struggled to penetrate. At the same time, IBM was able to offer its customers access to Apple's popular iOS devices, which were becoming more prevalent in the workplace.

However, Apple and IBM did not ignore their competition. They studied their rivals, such as Google and Microsoft, and recognized that they needed to differentiate their offering by providing customized, industry-specific solutions that catered to the unique needs of enterprise customers.

Through strategic prowess, Apple and IBM successfully executed their collaboration and developed a suite of enterprise-focused mobile apps and services, including apps for banking, retail, healthcare, and transportation industries. This collaboration between two tech giants showcases how embracing allies, studying rivals, and executing with strategic prowess can result in a successful competitive strategy.

6
ETHICAL STRATEGIC DECEPTION

Maxim V: *Effective competition requires strategic deception. To win, project weakness when strong, appear idle when active, feign distance when close, and simulate proximity when distant.*

Maxim of Competitive Wisdom

The thinking behind this wisdom is based on the belief that successful competition in business often involves deceiving one's competitors in order to gain an advantage. The idea is that by projecting weakness when actually strong, appearing idle when actually active, feigning distance when actually close, and simulating proximity when actually distant, a business can create a false sense of security or confusion among its competitors. This can lead the competitors to underestimate the business, misjudge its intentions, or make strategic errors, which in turn can create opportunities for the business to gain an advantage.

However, it is important to note that this type of strategic deception must be used carefully and ethically, as unethical or illegal tactics can harm a business's reputation and result in legal consequences. Additionally, while strategic deception can be a powerful tool in some cases, it is not always necessary or appropriate. Ultimately, the decision to use strategic deception should be based on a careful assessment of the specific situation and the potential risks and benefits involved.

THE ULTIMATE COMPETITIVE WISDOM

Real-Life Scenario
Apple Vs. Microsoft: Success through Strategic Deception

One real-life example that aligns with the competitive wisdom of strategic deception is the rivalry between technology giants Apple and Microsoft in the early 2000s.

During that time, Microsoft dominated the personal computer market with its Windows operating system, while Apple was struggling to regain its footing and market share. In an effort to regain its competitive edge, Apple employed a strategy of strategic deception.

Apple projected weakness when it was actually developing innovative products behind the scenes. The company released a series of lackluster products and faced financial difficulties, which led many to believe that Apple was on the verge of collapse. However, behind closed doors, Apple was secretly working on the development of game-changing devices such as the iPod, iPhone, and iPad.

Apple also appeared idle when it was actively pursuing new markets and technologies. While Microsoft was busy focusing on its core Windows operating system, Apple strategically shifted its attention to the development of portable music players, smartphones, and tablet computers. By seemingly ignoring Microsoft's dominance in the PC market, Apple successfully diverted attention and resources away from their true intentions.

Additionally, Apple feigned distance when it was actually in close competition with Microsoft. By positioning itself as an alternative and niche brand, Apple created a perception of being fundamentally different from its rival. This allowed Apple to attract a loyal customer base that valued Apple's unique design aesthetic and user experience, giving it a distinct competitive advantage.

THE ULTIMATE COMPETITIVE WISDOM

Lastly, Apple simulated proximity when it was actually distant from its competitors. Through clever marketing and strategic partnerships, Apple created an image of being at the forefront of technology and innovation. By associating itself with cutting-edge design and aligning with influential figures, Apple created the perception of being in close proximity to its customers while maintaining a significant distance from its competitors.

Overall, Apple's strategic use of deception allowed the company to effectively compete with Microsoft and eventually surpass it in terms of market capitalization and influence. The company's ability to project weakness when strong, appear idle when active, feign distance when close, and simulate proximity when distant exemplifies the competitive wisdom of strategic deception.

7
STRATEGIC UNPREDICTABILITY

Maxim VI: *Gain a competitive edge through strategic unpredictability. Diverge from their strengths, erode their confidence, leverage their weaknesses, and triumph with unforeseen approaches.*

Maxim of Competitive Wisdom

This competitive wisdom suggests that to gain an advantage over your competitors, you should employ strategic unpredictability in your approach. This means that you should be unpredictable in the way you operate and the decisions you make, to keep your competitors guessing and unsure of what you'll do next.

To achieve this, the wisdom advises that you should diverge from your competitor's strengths. By not following their lead and instead going in a different direction, you can make it harder for them to keep up with you and predict your moves.

Additionally, the wisdom suggests that you should erode your competitor's confidence. This can be achieved through various means, such as creating doubt in their abilities or highlighting weaknesses in their strategy. By doing this, you can make them more cautious and unsure of themselves, which can give you an advantage.

Furthermore, the wisdom recommends that you leverage your competitor's weaknesses. By identifying their weaknesses and using them to your advantage, you can put them on the back foot and gain an edge.

Finally, this wisdom advises that you should triumph with unforeseen approaches. This means that you should come up with

THE ULTIMATE COMPETITIVE WISDOM

creative and innovative ideas that your competitors won't expect. By doing so, you can gain an advantage that they can't match.

Overall, this competitive wisdom is about being strategic and adaptable in your approach to gain an advantage over your competitors. By being unpredictable, leveraging weaknesses, and triumphing with unforeseen approaches, you can stay ahead of the competition and achieve success in your endeavors.

Real-Life Scenario
Netflix Vs. Blockbuster

One real-life example of gaining a competitive edge through strategic unpredictability is the case of Netflix and its disruption of the traditional video rental industry dominated by companies like Blockbuster.

In the early 2000s, Blockbuster was the industry leader, with a vast network of brick-and-mortar stores offering physical movie rentals. At that time, the concept of streaming movies online was still in its infancy. However, Netflix, led by CEO Reed Hastings, recognized the potential of digital distribution and the changing preferences of consumers.

Netflix diverged from Blockbuster's strength of physical rentals and instead focused on leveraging the power of the internet. They introduced a subscription-based model that allowed customers to stream movies and TV shows directly to their devices. This unexpected approach disrupted the industry by offering convenience, affordability, and a wide range of content choices.

Netflix also eroded Blockbuster's confidence by challenging their late fee policy, which was a significant revenue source for Blockbuster. By eliminating late fees and adopting a no-due-dates policy, Netflix addressed a common pain point for customers and further differentiated themselves from the traditional rental experience.

THE ULTIMATE COMPETITIVE WISDOM

Furthermore, Netflix leveraged Blockbuster's weaknesses. The physical stores of Blockbuster had limited shelf space, leading to a limited selection of movies available for rental. Netflix, on the other hand, offered an extensive digital library that could be accessed from anywhere at any time.

The strategic unpredictability of Netflix, combined with their ability to diverge from the strengths of the incumbent industry leader, eroded their confidence, and leverage their weaknesses, led to their triumph in the market. Today, Blockbuster is a mere memory, while Netflix has become a global entertainment powerhouse, dominating the streaming industry, and revolutionizing how people consume media.

This example demonstrates the power of strategic unpredictability and the importance of exploring new avenues, challenging the status quo, and capitalizing on emerging technologies and changing consumer preferences to gain a competitive edge.

8
LOYAL WORKFORCE COLLABORATION

Maxim VII: *Treat your team like family, and they'll brave any challenge with you. See them as cherished kin, and they'll remain fiercely loyal through thick and thin.*

Maxim of Competitive Wisdom

This wisdom highlights the importance of treating your team with respect, care, and consideration, which can lead to increased loyalty, dedication, and commitment from your team members. By creating a family-like atmosphere, you can foster a sense of belonging, community, and mutual support that can help your team to overcome obstacles and work together more effectively.

When you treat your team as if they were your own children or beloved relatives, you communicate to them that you value their well-being, growth, and success, and that you're invested in their personal and professional development. This type of leadership style can inspire your team members to go above and beyond their normal duties, take ownership of their work, and work collaboratively towards shared goals.

Furthermore, when you show genuine care and concern for your team, they're more likely to reciprocate that attitude towards you and each other, creating a positive feedback loop that can strengthen the team dynamic. By contrast, if you treat your team members as disposable assets, or fail to recognize their contributions or struggles, you risk alienating them and decreasing their motivation and morale.

In short, the wisdom implies that by treating your team with respect, care, and empathy, you can create a strong foundation for a successful and competitive business, one where everyone is working together towards a shared goal and feels supported, valued, and motivated to contribute their best efforts.

Real-Life Scenario
Starbucks Family-like Culture

One real-life example of this wisdom can be observed in the leadership style of Howard Schultz, the former CEO of Starbucks. Schultz is known for treating his employees as family, and this approach has contributed significantly to the company's success.

Schultz believed that creating a culture of belonging and inclusivity was essential to building a successful business. He saw his employees not just as workers, but as cherished members of the Starbucks family. He made it a priority to connect with his employees on a personal level, listening to their concerns and ideas and empowering them to make a meaningful contribution to the company.

One example of Schultz's family-like approach to leadership can be seen in how Starbucks responded to the economic crisis of 2008. While other companies were laying off employees, Schultz made the decision to prioritize the well-being of his staff, even if it meant sacrificing profits. He offered healthcare benefits to all employees, including part-timers, and implemented a program that allowed workers to take online college courses for free.

By treating his employees like family, Schultz created a fiercely loyal workforce that was willing to go above and beyond to help the company succeed. Starbucks employees have been known to work long hours, take on extra shifts, and even make personal sacrifices to ensure that the company thrives.

This approach has paid off in many ways for Starbucks. The company has a low turnover rate and a reputation for excellent customer service, both of which have contributed to its success. Schultz's leadership style has become a case study in how treating employees like family can lead to a more engaged and motivated workforce and ultimately, a more successful business.

9
CHAOS CONCEALS OPPORTUNITY

Maxim VIII: *Chaos conceals opportunity, for those with the wisdom to find it.*

Maxim of Competitive Wisdom

The phrase "chaos conceals opportunity" is a competitive wisdom that suggests that in times of confusion, uncertainty, or disorder, there are opportunities that can be found by those with the wisdom and foresight to see them.

In business and competition, it's not uncommon to face situations that feel like chaos. New technologies and market shifts can disrupt established industries, leaving traditional players scrambling to adapt. Similarly, global events such as natural disasters, econcmic downturns, or political instability can create widespread uncertainty and unpredictability.

However, amidst the chaos, there are often opportunities to innovate, pivot, or gain an advantage over competitors. For example, a company that can quickly adapt to changing market conditions and consumer needs may be able to outmaneuver slower-moving rivals. Similarly, a business that can find a way to capitalize on a disruption or crisis may be able to gain market share or establish a new competitive edge.

The second part of the wisdom - "for those with the wisdom to find it" - is a reminder that identifying and seizing opportunities in chaotic situations requires a combination of creativity, strategic thinking, and the ability to see beyond immediate challenges. It's not enough to simply react to chaos or wait for things to settle down. Instead, successful competitors must be able to stay nimble, keep their eyes open for potential opportunities, and be willing to take calculated risks in order to capitalize on them.

THE ULTIMATE COMPETITIVE WISDOM

In summary, the wisdom that "chaos conceals opportunity, for those with the wisdom to find it" emphasizes the importance of remaining agile and adaptable in times of upheaval. By embracing change and uncertainty, and staying vigilant for potential opportunities, competitive players can find ways to thrive even in the midst of chaos.

Real-Life Scenario
The Story of 3M Post-it Notes

One real-life example of this wisdom is the story of how 3M discovered a highly profitable product by turning chaos into opportunity.

In the 1960s, a 3M researcher named Spencer Silver was trying to create a strong adhesive, but instead, he accidentally invented a weak one. The adhesive didn't stick very well and wasn't useful for anything, so Silver considered it a failure.

However, a few years later, another 3M employee named Art Fry was frustrated with his bookmarks falling out of his hymnal while singing in church. Remembering Silver's weak adhesive, he used it to coat his bookmarks and found that they stayed in place without damaging the pages.

3M saw the potential for this weak adhesive to be used in a variety of products and eventually created the now-famous Post-it Notes. The product was initially seen as a failure, but by recognizing the opportunity in the chaos of a failed experiment, 3M was able to turn it into a highly profitable product that is still widely used today.

This example demonstrates that chaos and failure can be an opportunity for growth and innovation if you have the wisdom and creativity to find it. 3M could have easily dismissed the weak adhesive as useless, but instead, they were able to turn it into a game-changing product.

THE ULTIMATE COMPETITIVE WISDOM

10

HARMONIZING OFFENSE AND DEFENSE

Maxim IX: *The path to invincibility is through a strong defense, while victory is achieved through a strategic offense.*

Maxim of Competitive Wisdom

This competitive wisdom is about balancing two important aspects of competing: defense and offense. In order to understand this wisdom, it is important to first define what defense and offense mean in a competitive context.

Defense refers to actions taken to protect oneself from an opponent's attacks or strategies. This could involve building up a strong barrier or fortification to prevent an opponent from breaking through or creating a system to detect and counter an opponent's actions. In a business context, defense could involve developing strong intellectual property protections, securing key partnerships or resources, or maintaining high-quality products or services that are difficult for competitors to replicate.

Offense, on the other hand, refers to actions taken to actively advance one's own goals or objectives. This could involve developing innovative new products or services, aggressively pursuing market share, or developing marketing campaigns to increase brand awareness and customer loyalty.

So, how do defense and offense relate to the idea of invincibility and victory?

The path to invincibility is through a strong defense because a solid defense can prevent an opponent from gaining ground or achieving their objectives. By creating a strong barrier or system to detect and counter an opponent's actions, one can minimize the risk of being caught off guard or losing ground to an opponent. In other words, a strong defense can create a sense of stability and security that allows one to focus on their own goals and objectives.

However, simply having a strong defense is not enough to achieve victory. Victory is achieved through a strategic offense because it is only by actively pursuing one's goals and objectives that one can gain ground and win the competition. By developing innovative new products or services, aggressively pursuing market share, or developing marketing campaigns to increase brand awareness and customer loyalty, one can gain an advantage over their competitors and achieve victory.

In summary, the path to invincibility is through a strong defense because it provides stability and security, while victory is achieved through a strategic offense because it allows one to actively pursue their goals and gain ground over competitors. Balancing both defense and offense is crucial for long-term success in any competitive environment.

Real-Life Scenario
Technology Industry Strategies: Lessons from Google and Amazon

One modern example of this wisdom can be seen in the technology industry, specifically in the case of cybersecurity. Cyber-attacks have become increasingly common and sophisticated, making it crucial for companies to have a strong defense in place to protect their data and systems.

One company that has demonstrated a strong defense strategy is Google. Google invests heavily in cybersecurity measures, including advanced encryption and multi-factor authentication. Additionally, Google's Security team employs hundreds of experts who work to identify and mitigate potential vulnerabilities in their systems.

THE ULTIMATE COMPETITIVE WISDOM

On the other hand, victory in the technology industry is often achieved through a strategic offense. A prime example of this is Amazon's approach to dominating the e-commerce market. Amazon consistently launches new products and services to stay ahead of the competition, including their popular Prime subscription service, same-day delivery, and their acquisition of Whole Foods.

However, Amazon has also faced several cyber-attacks and security breaches in the past. While their offense strategy has helped them achieve great success, they have also had to continually strengthen their defense to protect against potential cyber threats.

In conclusion, the technology industry is an example of how a strong defense is necessary to achieve invincibility, while a strategic offense is crucial for achieving victory. Companies like Google and Amazon demonstrate the importance of investing in both defense and offense strategies to stay ahead in a highly competitive market.

11
POWER OF DESIGN THINKING INNOVATION

Maxim X: *Limitations in resources do not hinder creativity or innovation. Strategic combinations of available resources can produce endless opportunities and possibilities, just like musical notes form new melodies.*

Maxim of Competitive Wisdom

This competitive wisdom highlights the power of creativity and innovation when resources are limited. Rather than viewing limitations as barriers, individuals and organizations should see them as opportunities for exploration. The analogy of musical notes forming melodies underscores the importance of strategic thinking in leveraging available resources. Just as a composer selects notes and their sequence to create a melody, individuals and organizations must strategically combine resources to achieve their goals. They need to be inventive in finding new ways to utilize what they have, fostering new opportunities and possibilities that may not have been feasible otherwise.

In addition to strategic and creative thinking, empathy plays a vital role when facing resource limitations. Design thinking emphasizes understanding the needs and desires of the target audience. By incorporating empathy into strategic thinking, individuals and organizations gain a deeper understanding of their users' challenges, aspirations, and preferences. This understanding enables them to tailor resource combinations and innovative solutions to better meet user needs. Like a composer considering the emotions and preferences of their audience, individuals and organizations can leverage empathy to design solutions that resonate with their target audience. By empathizing with users and considering their unique perspectives, they can create opportunities and possibilities that genuinely address their audience's needs, even in the face of resource limitations.

Real-Life Scenario
Airbnb: Revolutionizing Hospitality Through Design Thinking

One real-life example of a global business that demonstrates the application of this design thinking is Airbnb.

Airbnb is a home-sharing platform that was founded in 2008 when its founders realized that they did not have enough money to pay rent for their apartment. Instead of giving up, they found a creative solution by renting out air mattresses in their living room to conference attendees who were looking for affordable accommodations. This inspired them to create a website where people could rent out their homes to travelers, and Airbnb was born.

Despite being a small startup with limited resources at the time, Airbnb leveraged the power of strategic combinations of available resources to create a new and innovative business model. They combined the growing popularity of the sharing economy, the ubiquity of smartphones, and the power of social media to create a platform that disrupted the traditional hotel industry.

Today, Airbnb is a global brand valued at over $100 billion. They have over 4 million listings worldwide and have hosted over 800 million guests. Airbnb's success is a testament to the power of creativity and innovation, and how strategic combinations of available resources can produce endless opportunities and possibilities, just like musical notes form new melodies.

12
GREATEST COMPETITIVE EDGE

Maxim XI: *Realizing our full potential unleashes our greatest competitive edge.*

Maxim of Competitive Wisdom

The thinking behind the competitive wisdom "Realizing our full potential unleashes our greatest competitive edge" is that individuals or organizations can achieve their best performance by tapping into their full potential. This means understanding and utilizing all their skills, abilities, and resources to the best of their ability.

In a competitive environment, where businesses, organizations, or individuals are striving to outperform each other, those who realize their full potential are likely to have a significant advantage. When individuals or organizations operate at their best, they can offer unique value propositions to their customers, innovate more effectively, make better decisions, and ultimately achieve greater success.

However, realizing our full potential requires a deep understanding of ourselves, our strengths, weaknesses, and how we can leverage them. It also requires effort, dedication, and a willingness to continuously improve and learn.

In summary, realizing our full potential is crucial to achieving success in a competitive environment. By tapping into our full potential, we can unlock our greatest competitive edge, outperform our rivals, and achieve our goals.

THE ULTIMATE COMPETITIVE WISDOM

Real-Life Scenario
The Michael Jordan Story

A real-life example of the application of the thinking that realizing our full potential unleashes our greatest competitive edge is the story of Michael Jordan, widely considered the greatest basketball player of all time.

Jordan was known for his tireless work ethic and his commitment to constantly improving his skills and abilities on the court. He famously practiced for hours every day, long after his teammates had gone home, and pushed himself to his physical and mental limits.

Jordan's relentless pursuit of excellence allowed him to dominate the game of basketball for over a decade, winning six NBA championships, five MVP awards, and numerous other accolades. His dedication to realizing his full potential not only made him an unstoppable force on the court but also inspired his teammates and opponents alike to raise their own game.

By pushing himself to be the best he could be, Jordan set a new standard of excellence in the sport and cemented his place in basketball history. His example serves as a powerful reminder that realizing our full potential can be our greatest competitive edge, allowing us to achieve extraordinary success in any field we choose to pursue.

13
PREPARATION AND DECISIVENESS

Maxim XII: *Victory is achieved before the competition begins. Success belongs to those who prepare meticulously and act decisively, while failure awaits those who enter the arena unprepared.*

Maxim of Competitive Wisdom

This competitive wisdom emphasizes the importance of preparation and decisiveness in achieving success. It suggests that the outcome of a competition is determined before it even begins, and victory is only attainable for those who have prepared well and taken decisive actions.

The first part of the wisdom states that "victory is achieved before the competition begins." This means that success is not determined solely by the performance during the competition itself but is largely influenced by the preparation that was done beforehand. The preparation can take many forms, such as conducting market research, developing a strategic plan, building a skilled team, or training extensively. The preparation ensures that the competitors have a strong foundation to build upon and a clear direction to follow.

The second part of the wisdom states that "success belongs to those who prepare meticulously and act decisively." This highlights the importance of both preparation and action. Preparation alone is not enough to guarantee success; it must be followed by decisive action. In a competitive environment, quick and effective decision-making is crucial, and hesitation or indecision can result in missed opportunities or even failure.

THE ULTIMATE COMPETITIVE WISDOM

Finally, the wisdom warns that "failure awaits those who enter the arena unprepared." This means that those who enter a competition without proper preparation and planning are more likely to fail. Without a clear strategy or a solid foundation, competitors may struggle to adapt to changing circumstances or to overcome unexpected challenges.

In summary, this competitive wisdom emphasizes the importance of preparation, decisiveness, and action in achieving success. It suggests that victory is achieved before the competition begins, and those who prepare meticulously, and act decisively are more likely to succeed, while those who enter the arena unprepared are more likely to fail.

Real-Life Scenario
Amazon's Disruptive Success

One example of the application of this thinking is the success story of the online retailer Amazon. Before Amazon entered the market, traditional retailers dominated the industry, and the idea of purchasing products online was still in its infancy. However, Amazon's founder, Jeff Bezos, understood the potential of the internet to transform the retail industry and saw an opportunity to disrupt the traditional model.

Bezos meticulously prepared for the launch of Amazon, conducting extensive market research to identify the most promising products and studying the behavior of online shoppers. He also created a business model that focused on providing customers with a vast selection, competitive prices, and exceptional customer service.

Amazon's success was not just due to its innovative business model, but also its ability to act decisively. Bezos made strategic investments in technology, logistics, and marketing to grow the company quickly and maintain a competitive edge. He also

THE ULTIMATE COMPETITIVE WISDOM

recognized the importance of continuous improvement and invested heavily in research and development to stay ahead of the competition.

By knowing the market and himself, and by preparing meticulously and acting decisively, Jeff Bezos was able to achieve victory in the highly competitive retail industry. Today, Amazon is the world's largest online retailer, with a market capitalization of over $1 trillion, and its success has transformed the entire retail industry.

14
OUTMANEUVERING YOUR COMPETITION

Maxim XIII: *Winning in competition isn't about direct confrontation, but rather weakening your opponent's resolve through effective strategy and execution.*

Maxim of Competitive Wisdom

This competitive wisdom is rooted in the idea that competition is not only about brute force or direct confrontation but also about outmaneuvering your opponent. It suggests that success in competition is not determined by who has the most strength or firepower, but by who has the most effective strategy and execution.

In a competitive environment, direct confrontation can be costly and risky, and often results in a lose-lose scenario for both parties. Instead, a more effective approach is to weaken the opponent's resolve and create an advantage by utilizing effective strategy and execution.

Effective strategy involves analyzing your opponent's strengths and weaknesses and finding ways to capitalize on them. By understanding your opponent's vulnerabilities, you can devise a plan to exploit them, leading to an advantage.

Execution is the key to any successful strategy. Even the best strategy can fail if not executed properly. Proper execution requires discipline, focus, and attention to detail. It's about making sure that every action is in line with the overall strategy, and that every opportunity is seized to gain an advantage.

By adopting this competitive wisdom, individuals and organizations can gain a competitive edge without engaging in direct confrontation. They can identify and exploit their opponent's vulnerabilities while focusing on their own strengths, leading to a more successful outcome in the competition.

Real-Life Scenario
Nike vs. Reebok: Strategic Maneuvers in the Sportswear Rivalry

Nike and Reebok have long been competitors in the athletic footwear and apparel market. In this rivalry, winning has often been about more than just direct confrontation; it has been about weakening the opponent's resolve through effective strategy and execution.

Nike, known for its innovative products and powerful branding, has consistently sought to outmaneuver Reebok and gain a competitive edge. One notable example is Nike's strategy of securing endorsements from high-profile athletes. By signing endorsement deals with legendary sports figures like Michael Jordan, Serena Williams, and LeBron James, Nike not only gained credibility but also weakened Reebok's ability to secure similar endorsements. Nike's strong athlete partnerships and iconic "swoosh" logo have helped solidify its position as a market leader and have had a significant impact on its brand perception.

In response, Reebok has employed its own strategic moves to counter Nike's dominance. One notable example is Reebok's focus on targeted niches within the market. Recognizing that it couldn't directly compete with Nike in terms of market share, Reebok strategically positioned itself as a brand catering to specific sports or fitness segments. For instance, Reebok gained popularity in the CrossFit community by designing shoes and apparel specifically tailored for CrossFit athletes. By carving out these niches, Reebok aimed to weaken Nike's stronghold on the broader athletic market and attract a dedicated customer base.

THE ULTIMATE COMPETITIVE WISDOM

The Nike vs. Reebok rivalry demonstrates that winning in competition often involves more than direct confrontation. Both companies have utilized strategic maneuvers to weaken each other's resolve and gain an advantage. Nike's focus on high-profile endorsements and powerful branding has helped it maintain a dominant position, while Reebok's niche targeting has allowed it to establish a foothold in specific segments. Through effective strategy and execution, both companies have engaged in a battle for market share and consumer loyalty in the sportswear industry.

15
STRATEGIC RECONNAISSANCE

Maxim XIV: *The visionary leader and astute strategist harness the collective intelligence of their team, employing it for strategic reconnaissance, ultimately achieving extraordinary outcomes.*

Maxim of Competitive Wisdom

The thinking behind this competitive wisdom emphasizes the importance of a visionary leader and an astute strategist in leveraging the collective intelligence of their team to achieve exceptional results through strategic reconnaissance.

A visionary leader possesses a clear understanding of the organization's mission, goals, and long-term vision. They have the ability to inspire and guide their team towards a common purpose, aligning individual efforts with the overall strategic direction. Such a leader fosters a culture of innovation, encourages creative thinking, and values diverse perspectives within the team.

An astute strategist, on the other hand, possesses a deep understanding of the competitive landscape, market trends, and the organization's strengths and weaknesses. They analyze and interpret information, both internal and external, to identify opportunities and potential threats. By synthesizing this information, they formulate effective strategies that give the organization a competitive advantage.

The concept of harnessing collective intelligence highlights the importance of teamwork and collaboration. A visionary leader recognizes the value of tapping into the diverse skills, knowledge, and experiences of their team members. They create an environment that encourages open communication, idea-sharing, and constructive debate. By leveraging the collective intelligence of the team, the leader gains access to a wider range of perspectives and ideas, which can greatly enhance the quality of strategic decision-making.

Strategic reconnaissance refers to the process of gathering and analyzing information about the competitive landscape, market conditions, and the organization's internal capabilities. It involves conducting thorough research, monitoring industry trends, evaluating competitor actions, and assessing the organization's own strengths and weaknesses. This reconnaissance allows the leader and strategist to make informed decisions, identify potential opportunities and threats, and devise effective strategies accordingly.

The ultimate goal of this thinking is to achieve extraordinary outcomes. By combining visionary leadership, astute strategic thinking, and harnessing collective intelligence, organizations can gain a competitive edge, seize opportunities, and overcome challenges. This approach enables them to make well-informed decisions, adapt to changing circumstances, and outperform competitors.

Through effective strategic reconnaissance, the leader and strategist gain valuable insights into the market, enabling them to anticipate trends, identify gaps in the competition, and align the organization's resources and capabilities accordingly. This proactive approach helps the organization stay ahead of the curve and position itself for success.

Overall, the thinking behind this competitive wisdom recognizes the significance of a visionary leader and astute strategist who are capable of harnessing the collective intelligence of their team. By employing strategic reconnaissance and leveraging the diverse perspectives within the organization, they can make informed decisions, formulate effective strategies, and achieve extraordinary outcomes in the highly competitive business landscape.

THE ULTIMATE COMPETITIVE WISDOM

Real-Life Scenario
Harnessing Collective Intelligence for Global Climate Action

The Paris Climate Agreement, adopted in 2015, serves as a prime example of harnessing collective intelligence and strategic reconnaissance to address the global challenge of climate change.

In response to the urgent need to combat climate change, leaders from nearly 200 countries came together to negotiate and develop a comprehensive agreement. The collective intelligence of scientists, environmental experts, and policymakers played a crucial role in understanding the risks associated with climate change and identifying effective strategies for mitigation and adaptation.

Prior to the agreement, extensive research and data analysis were conducted by scientific bodies such as the Intergovernmental Panel on Climate Change (IPCC). This collective intelligence informed the negotiations, providing evidence of the impacts of climate change, the necessary emission reduction targets, and the importance of international cooperation.

During the negotiations, strategic reconnaissance was employed to understand the diverse perspectives, interests, and priorities of participating nations. Intensive discussions and diplomatic efforts allowed countries to bridge differences and find common ground. The agreement's framework emphasized a collective response, recognizing the principle of common but differentiated responsibilities among developed and developing nations.

The Paris Climate Agreement established a global commitment to limit global warming well below 2 degrees Celsius above pre-industrial levels and to pursue efforts to limit the temperature increase to 1.5 degrees Celsius. It also called for countries to regularly report on their emission reduction efforts, adapt to the

THE ULTIMATE COMPETITIVE WISDOM

impacts of climate change, and provide financial and technological support to developing nations.

The agreement's success lies in its ability to unite nations around a shared vision and set a framework for global climate action. It represents a significant achievement in leveraging collective intelligence to address the complex and interconnected issue of climate change.

Since its adoption, the Paris Climate Agreement has prompted increased ambition and action on climate change, with countries revising and strengthening their emission reduction targets. It has also fostered global collaboration in areas such as renewable energy, sustainable development, and climate finance.

The Paris Climate Agreement exemplifies the competitive wisdom of harnessing collective intelligence and strategic reconnaissance to tackle pressing geopolitical challenges. It showcases the power of international cooperation and the capacity of nations to come together, using shared knowledge and insights, to address complex global issues such as climate change.

16
COMPLACENCY ERODES EXCELLENCE

Maxim XV: *Complacency erodes excellence, leaving greatness to wither.*

Maxim of the Competitive Wisdom

The thinking behind the competitive wisdom "Complacency erodes excellence, leaving greatness to wither" is that when individuals, organizations or teams become too comfortable with their current level of success, they may start to neglect the hard work, innovation, and attention to detail that helped them achieve that success in the first place. This can lead to a decline in performance, eroding the excellence that was once achieved, and eventually resulting in a loss of the competitive edge that made the individual or organization great.

Complacency can arise from a number of factors, including success, routine, and comfort. When an individual or organization achieves a certain level of success, they may become overconfident, believing that they have all the answers and that their methods and processes are infallible. This overconfidence can lead to a lack of awareness of the external environment, including changes in the market, new competitors, and technological advancements.

Similarly, when individuals or organizations become too comfortable in their routine or are unwilling to change, they may miss out on new opportunities for growth and innovation. In a rapidly changing world, standing still is equivalent to moving backwards, and complacency can quickly erode the competitive advantage that was once enjoyed.

THE ULTIMATE COMPETITIVE WISDOM

Excellence requires continuous effort and a dedication to improvement. In order to remain competitive, individuals and organizations must constantly challenge themselves to do better, think creatively, and strive for excellence in all areas. This involves embracing change, taking calculated risks, and being open to new ideas and approaches.

Ultimately, the thinking behind "Complacency erodes excellence, leaving greatness to wither" is a reminder that success is not permanent, and that the pursuit of excellence requires a continuous effort to adapt, innovate, and push boundaries. Those who become complacent risk falling behind, losing their competitive edge, and seeing their greatness wither away.

Real-Life Scenario
The Downfall of Kodak

One real-life example for the application of this thinking is the downfall of Kodak, a once-dominant player in the photography industry.

Kodak, in its prime, held a strong position in the market, with its film-based cameras and film processing services being widely popular. However, the company failed to adapt and understand the shifting dynamics of the industry it operated in.

With the advent of digital photography, Kodak faced a new breed of competitors who embraced the technological advancements. Despite being aware of the rising popularity of digital photography, Kodak was complacent and failed to acknowledge the disruptive potential it held.

Instead of leveraging their expertise and resources to innovate and embrace the digital revolution, Kodak remained focused on their film-based business. They underestimated the growing demand for digital cameras and the convenience they offered to consumers.

THE ULTIMATE COMPETITIVE WISDOM

This lack of understanding and complacency proved detrimental to Kodak's long-term success. While they knew themselves as a film-based photography company, they failed to fully comprehend the changing needs and preferences of their customers or acknowledge the fierce competition emerging in the digital photography space.

Consequently, Kodak's market share and profitability declined rapidly, leading to their eventual bankruptcy filing in 2012. Meanwhile, companies like Canon, Nikon, and later smartphone manufacturers like Apple and Samsung, who understood the changing landscape and adapted accordingly, flourished.

This real-life example demonstrates the importance of knowing both the self (Kodak's film-based expertise) and the enemy (the rising digital photography trend) in order to avoid complacency and maintain excellence. By neglecting to understand the competitor and failing to adapt, even a once-great company like Kodak found itself withering away.

17
SYNERGY OF STRATEGY AND TACTICS

Maxim XVI *The synergy of strategy and tactics is the cornerstone of triumph; one without the other leads to failure.*

Maxim of Competitive Wisdom

The competitive wisdom of "The synergy of strategy and tactics is the cornerstone of triumph; one without the other leads to failure" highlights the crucial relationship between strategy and tactics in achieving success in any competitive endeavor. It emphasizes that strategy and tactics are two interconnected elements that work together harmoniously to achieve a desired outcome.

Strategy refers to an overarching plan or approach designed to achieve long-term goals. It involves analyzing the competitive landscape, understanding the strengths and weaknesses of both oneself and the competition, and formulating a comprehensive plan to gain a competitive advantage. Strategy provides a framework and direction for decision-making, resource allocation, and setting priorities.

On the other hand, tactics are the specific actions and maneuvers implemented within the strategic framework to execute the plan effectively. Tactics are more short-term and focused on the immediate objectives, taking into account the dynamic nature of the situation and adapting accordingly. They involve the deployment of resources, operational maneuvers, and exploiting opportunities to gain an advantage over opponents.

The wisdom highlights that strategy and tactics are inseparable and interdependent. Strategy provides the big picture vision and guides the allocation of resources, while tactics bring the strategy to life through execution and adaptability. Strategy without tactics can lead to a lack of implementation, leaving a plan ineffective and incomplete. Similarly, tactics without a strategic foundation lack direction and may result in scattered efforts that fail to achieve the desired objectives.

To achieve triumph, it is crucial to synchronize strategy and tactics effectively. Strategy provides the overarching framework that sets the direction and goals, while tactics execute the strategy in a flexible and dynamic manner, adapting to changing circumstances. The synergy between strategy and tactics ensures that actions are aligned with the overall vision and objectives, optimizing resource allocation, and maximizing the chances of success.

In summary, this competitive wisdom emphasizes the critical importance of strategy and tactics working together. It underscores that a successful outcome is achieved when strategy provides the roadmap and tactics execute the plan with agility and adaptability. Neglecting either element can lead to failure, reinforcing the need for a holistic and integrated approach to competitive endeavors.

Real-Life Scenario
Operation Desert Storm: Triumph Through Strategy and Tactics

During the Gulf War in 1990-1991, the military campaign known as Operation Desert Storm provides a real-life example that exemplifies the competitive wisdom stating that "the synergy of strategy and tactics is the cornerstone of triumph; one without the other leads to failure."

In response to Iraq's invasion of Kuwait, a coalition of international forces led by the United States embarked on a mission to liberate Kuwait and diminish the threat posed by Saddam Hussein's regime. This operation involved complex

THE ULTIMATE COMPETITIVE WISDOM

military planning, coordination, and the integration of strategy and tactics.

The coalition forces, under the leadership of General Norman Schwarzkopf, meticulously developed a comprehensive strategy to achieve their objectives. The strategic plan aimed to rapidly weaken Iraq's military capabilities through a combination of aerial bombardment, ground assaults, and decisive maneuvering.

The tactical execution of this strategy was equally crucial to ensure success. The coalition forces employed a range of tactical approaches, including precision airstrikes, ground assaults, and specialized operations to disrupt enemy communication and supply lines. These tactics were carefully orchestrated to exploit vulnerabilities in Iraq's defenses while minimizing risks to the coalition forces.

The synergy between strategy and tactics became evident during the operation. The initial aerial campaign crippled Iraq's air defenses and communication infrastructure, establishing air superiority for the coalition forces. This paved the way for ground troops to launch a well-coordinated assault, combining armored units, infantry, and air support to swiftly push back Iraqi forces.

Throughout the conflict, the coalition forces demonstrated the ability to adapt their tactics to changing circumstances, leveraging technology, intelligence, and rapid decision-making. They effectively combined strategic objectives with tactical maneuvers, resulting in the successful liberation of Kuwait and the defeat of the Iraqi forces.

In contrast, Iraq's military strategy lacked the necessary coherence and integration of tactics. They were unable to counter the coalition's superior strategic planning and tactical execution, leading to their eventual defeat.

The example of Operation Desert Storm illustrates the significance of aligning strategy and tactics for triumph in competitive situations. It highlights how a well-crafted strategy, when

THE ULTIMATE COMPETITIVE WISDOM

implemented through effective tactical maneuvers, can lead to decisive victories. Conversely, an absence of synergy between strategy and tactics can hamper success and lead to failure.

By combining strategic vision with tactical expertise, the coalition forces in Operation Desert Storm demonstrated the power of coordinated planning, intelligent decision-making, and the seamless integration of strategy and tactics. Their triumph further emphasizes the importance of the synergy between these two elements in achieving success in competitive endeavors.

18
SACRIFICE AND SUCCESS

Maxim XVII: *Success favors those who embrace sacrifice.*

Maxim of Competitive Wisdom

The competitive wisdom behind the statement "Success favors those who embrace sacrifice" revolves around the understanding that achieving significant success often requires individuals to make sacrifices along the way. This wisdom recognizes that true accomplishments are not attained without dedication, hard work, and the willingness to give up certain comforts or immediate gratification.

When individuals embrace sacrifice, they demonstrate their commitment to their goals and aspirations. They understand that success rarely comes without effort, perseverance, and the willingness to make difficult choices. Sacrifices may come in various forms, such as time, leisure activities, personal relationships, financial investments, or even the relinquishment of certain immediate desires. It involves prioritizing long-term gains over short-term pleasures.

By willingly embracing sacrifice, individuals exhibit discipline, determination, and a strong work ethic. They recognize that achieving greatness requires stepping out of their comfort zones and pushing beyond perceived limitations. Sacrifices can involve investing countless hours into honing their skills, forgoing immediate gratification to stay focused on their goals, or making difficult decisions that may not be popular but are necessary for long-term success.

Moreover, embracing sacrifice also entails understanding that setbacks and failures are inevitable on the path to success. Sacrifices often involve taking risks and accepting that not all attempts will lead to immediate triumph. However, those who embrace sacrifice perceive setbacks as valuable learning experiences, opportunities for growth, and steppingstones towards ultimate success.

Ultimately, the wisdom behind "Success favors those who embrace sacrifice" suggests that those who are willing to put in the necessary effort, make difficult choices, and endure temporary discomfort are more likely to achieve their goals. By recognizing the significance of sacrifice and actively incorporating it into their journey, individuals increase their chances of attaining the success they aspire to, regardless of the challenges they may face or the obstacles that may come their way.

Real-Life Scenario
Sacrifice and Success: The Inspiring Journey of Oprah Winfrey

Oprah Winfrey, the renowned media mogul and philanthropist, serves as a compelling real-life example that aligns with the competitive wisdom of "success favors those who embrace sacrifice." From humble beginnings and overcoming adversity, Oprah's journey to success required significant sacrifices and unwavering commitment.

Born into economic hardship in rural Mississippi, Oprah faced numerous challenges throughout her childhood. However, she recognized the power of education and dedicated herself to excelling academically. Sacrificing time and energy, she pursued her passion for media and storytelling, beginning her career in radio and later transitioning to television.

THE ULTIMATE COMPETITIVE WISDOM

Despite encountering setbacks and obstacles along the way, Oprah remained steadfast in her pursuit of success. She willingly made sacrifices, such as leaving a successful talk show in Chicago to launch her own television network, OWN. Although the network initially faced financial difficulties, Oprah's determination, and willingness to sacrifice personal comfort propelled its eventual success.

Furthermore, Oprah's openness about her personal struggles, including childhood trauma and weight battles, showcased her authenticity and vulnerability. By sacrificing privacy and openly discussing her experiences, she connected with her audience on a deeper level and addressed important social issues. Through these actions, she inspired others and fostered a sense of empathy and empowerment.

Oprah's sacrifices, dedication, and unwavering work ethic have been instrumental in her remarkable success. Building a media empire and becoming a billionaire, she solidified her position as one of the most influential women in the world. Through her philanthropic endeavors, such as the Oprah Winfrey Leadership Academy for Girls and various charitable initiatives, she continues to make a positive impact and uplift others.

The example of Oprah Winfrey exemplifies the notion that embracing sacrifice, taking risks, and remaining committed to one's vision can lead to extraordinary achievements. Despite facing adversity and making personal sacrifices along the way, Oprah's unwavering determination and resilience have propelled her to unparalleled success and established her as an iconic figure in numerous fields.

19
BALANCING AGGRESSION AND PROTECTION

Maxim XVIII: *Unleash offense from celestial heights, shield defense in subterranean depths.*

Maxim of Competitive Wisdom

The competitive wisdom of "Unleash offense from celestial heights, shield defense in subterranean depths" emphasizes a strategic approach that combines both aggression and protection. Let's delve into the detailed thinking behind this statement:

Unleash offense from celestial heights: This phrase suggests adopting an aggressive and proactive mindset. Just like celestial heights represent an elevated position, it encourages taking advantage of your strengths, resources, and unique capabilities to launch powerful attacks. It implies seeking opportunities, being assertive, and leveraging your advantages to gain a competitive edge. By unleashing offense from celestial heights, you aim to dominate the battlefield, whether it be in business, sports, or any other competitive arena.

Shield defense in subterranean depths: While offense is essential, defense plays an equally crucial role in achieving long-term success. The metaphor of subterranean depths highlights the importance of a solid and well-protected defense. It signifies going beyond surface-level defense and establishing a robust system that is resilient, adaptable, and difficult to breach. By shielding defense in subterranean depths, you prioritize safeguarding your core assets, mitigating risks, and fortifying your position against potential threats or competitors' attacks.

THE ULTIMATE COMPETITIVE WISDOM

Together, these concepts represent a balanced approach to competition. By leveraging offense from celestial heights, you strive to gain an advantage and seize opportunities actively. Simultaneously, by reinforcing your defense in subterranean depths, you ensure the sustainability and longevity of your competitive position. It signifies a strategic mindset that combines calculated aggression with smart and resilient defense, leading to increased chances of success and the ability to weather challenges effectively.

Real-life Scenario
Dynamics of Modern Warfare

In the world of modern warfare, the United States Air Force's utilization of air power and its integration with ground-based missile defense systems exemplify the strategic concept of "unleashing offense from celestial heights, shielding defense in subterranean depths."

In terms of offense, the U.S. Air Force possesses a formidable aerial arsenal, including long-range bombers, fighter jets, and unmanned aerial vehicles (UAVs). These assets allow them to project power from the skies, striking targets with precision and overwhelming force. By leveraging their superiority in the air, they gain a distinct advantage over their adversaries, neutralizing threats and disrupting enemy operations. This offensive capability grants them the ability to engage in conflicts swiftly and decisively across the globe.

On the other hand, the subterranean depths represent the realm of missile defense systems. The United States has invested heavily in developing and deploying advanced missile defense technologies, such as the Ground-Based Midcourse Defense (GMD) system. These systems are strategically positioned in underground facilities, capable of detecting, tracking, and intercepting incoming ballistic missiles.

THE ULTIMATE COMPETITIVE WISDOM

By shielding their defense capabilities in subterranean depths, the U.S. effectively creates multiple layers of protection against potential missile threats. This layered defense approach minimizes vulnerabilities and increases the chances of intercepting enemy missiles before they can reach their intended targets. It acts as a crucial defensive shield, safeguarding vital assets, military installations, and even civilian populations from potential missile attacks.

The integration of offensive capabilities from celestial heights and defensive measures in subterranean depths allows the United States to maintain a comprehensive and formidable strategic posture. By possessing unrivaled air power and a robust missile defense system, they not only project strength and deter potential adversaries but also ensure their ability to respond effectively in the event of an attack.

This real-life example demonstrates the importance of balancing offensive and defensive capabilities in a competitive landscape. By leveraging their strengths in the air while fortifying their defenses, nations can establish a dominant position and minimize the risks they face in the ever-evolving dynamics of modern warfare.

20
ART OF MINIMAL FORCE

Maxim XIX: *Unleash the power of the few, conquer the impossible.*

Maxim of Competitive Wisdom

"Unleash the power of the few, conquer the impossible" encapsulates a strategic mindset that emphasizes the value of focused efforts and the ability to achieve remarkable results despite limited resources. It recognizes the potential for small, highly skilled teams or individuals to make a significant impact in competitive arenas.

By "unleashing the power of the few," the phrase encourages harnessing the collective strength and expertise of a select few rather than spreading resources thin across a large group. This approach prioritizes quality over quantity, focusing on assembling a team or individuals with exceptional abilities, knowledge, and dedication.

"Conquer the impossible" represents the ambitious nature of this competitive wisdom. It inspires a relentless pursuit of success and overcoming seemingly insurmountable challenges. Rather than being discouraged by daunting obstacles, the quote encourages embracing them as opportunities for innovation, creativity, and breakthroughs.

This strategic mindset is rooted in the understanding that success does not always depend on overwhelming force or extensive resources. It highlights the importance of strategic thinking, agility, and adaptability. It's about leveraging strengths, identifying weaknesses in the competition, and using limited resources efficiently to achieve extraordinary outcomes.

THE ULTIMATE COMPETITIVE WISDOM

Overall, the quote encourages a focused and determined approach to competition, where small but highly skilled forces can outmaneuver and outperform larger, less focused opponents. It celebrates the ability to defy conventional expectations, accomplish the seemingly impossible, and leave a lasting impact.

Real-Life Scenario
Apollo 11 Moon Landing: Conquering the Impossible

One real-life example for the application of the thinking "Unleash the power of the few, conquer the impossible" can be seen in the story of the Apollo 11 moon landing.

During the 1960s, the United States and the Soviet Union were engaged in the Space Race, competing to achieve significant milestones in space exploration. In 1961, President John F. Kennedy declared the ambitious goal of landing a man on the moon and returning him safely to Earth by the end of the decade.

To accomplish this seemingly impossible feat, NASA had to meticulously plan and execute a series of missions, each building upon the knowledge and experience gained from previous endeavors. The engineers and scientists at NASA were well aware of the challenges that lay ahead, including the technological complexities and the risks involved.

The success of the Apollo 11 mission hinged on the ability of NASA to leverage the expertise and skills of a select group of individuals who had an in-depth understanding of both the enemy (the harsh conditions and unknown variables of space) and themselves (their own capabilities and limitations).

The team at NASA conducted extensive research, testing, and simulations to gain a thorough understanding of the lunar environment and the necessary equipment and procedures. They also carefully selected and trained the astronauts, ensuring they possessed the necessary skills, knowledge, and physical fitness to handle the mission's demands.

By knowing the enemy (the challenges of space) and knowing themselves (their own capabilities and limitations), NASA was able to strategize, innovate, and overcome the obstacles standing in the way of landing humans on the moon. The culmination of their efforts was realized on July 20, 1969, when Neil Armstrong and Buzz Aldrin became the first humans to set foot on the lunar surface.

The Apollo 11 mission stands as a powerful example of how understanding the enemy (the harsh conditions of space) and knowing oneself (the capabilities of the team and astronauts) can lead to the conquest of what was once considered impossible. It demonstrates the potential to achieve remarkable feats by harnessing the power of a dedicated few who possess the necessary knowledge, skills, and determination.

21
ACHIEVING EXPONENTIAL SUCCESS

Maxim XX: *Seize opportunities, multiply success.*

Maxim of Competitive Wisdom

The thinking behind the competitive wisdom "Seize opportunities, multiply success" is rooted in the recognition that success is not solely dependent on hard work and talent but also on the ability to identify and capitalize on favorable circumstances. It emphasizes the importance of actively seeking out and seizing opportunities to achieve greater levels of success, rather than passively waiting for them to come.

At its core, this wisdom is based on the understanding that opportunities are not static; they arise and evolve in response to changing market dynamics, technological advancements, consumer needs, and other external factors. By actively scanning the environment and staying attuned to these shifts, individuals and organizations can position themselves to take advantage of emerging opportunities.

Seizing opportunities involves a combination of vision, creativity, and strategic thinking. It requires the ability to identify gaps in the market, anticipate trends, and envision innovative solutions. This thinking goes beyond mere observation and analysis; it involves actively seeking out ways to leverage strengths, resources, and capabilities to create new value.

Moreover, seizing opportunities often requires a willingness to take calculated risks. It involves stepping out of one's comfort zone, embracing uncertainty, and making bold moves. It requires the ability to overcome fear of failure and view setbacks as learning experiences rather than roadblocks. Successful individuals and organizations understand that growth and progress often involve taking risks and making strategic bets.

However, seizing opportunities alone is not enough. The concept of multiplying success highlights the importance of building upon initial achievements to create a compounding effect. It involves capitalizing on early wins, leveraging momentum, and continuously pushing the boundaries of what is possible. By multiplying success, individuals and organizations can expand their impact, unlock new opportunities, and sustain long-term growth.

To effectively apply this competitive wisdom, individuals and organizations need to cultivate a mindset of agility, adaptability, and continuous learning. They must be open to new ideas, embrace change, and stay ahead of the curve. It requires a proactive approach to innovation, constantly seeking ways to evolve and improve.

In summary, the thinking behind "Seize opportunities, multiply success" revolves around actively seeking and seizing opportunities, taking calculated risks, and building upon initial achievements to create a compounding effect. It requires a combination of vision, creativity, strategic thinking, adaptability, and a willingness to embrace change. By embodying this wisdom, individuals and organizations can position themselves for sustained success in a competitive landscape.

Real-Life Scenario
From Electric Cars to Space Exploration: Elon Musk's Journey of Seizing Opportunities

"Seize opportunities, multiply success" emphasizes the importance of recognizing and acting upon opportunities to achieve greater levels of success. By seizing opportunities, individuals and organizations can propel themselves forward and maximize their achievements.

For instance, consider the story of Elon Musk, the CEO of Tesla and SpaceX. Musk recognized the potential for electric vehicles to revolutionize the automotive industry and address environmental concerns. Seizing this opportunity, he founded Tesla with the goal of producing high-performance electric cars that could compete with traditional gasoline-powered vehicles. Despite facing skepticism and challenges, Musk persevered and pushed the boundaries of innovation in electric vehicle technology.

Tesla's success in the electric vehicle market opened doors for Musk to further multiply his achievements. Recognizing the need for sustainable space exploration, he founded SpaceX to reduce the costs associated with space travel and make it more accessible. SpaceX's groundbreaking achievements, such as reusable rockets and successful satellite launches, have not only transformed the aerospace industry but have also positioned Musk as a key player in the space race.

By seizing the opportunity in electric vehicles and then expanding into the aerospace industry, Musk exemplifies the concept of seizing opportunities to multiply success. His ability to recognize and act upon these opportunities has propelled him and his companies to unprecedented levels of achievement, disrupting industries and inspiring innovation.

The lesson from Musk's example is that seizing opportunities requires foresight, determination, and a willingness to take calculated risks. It involves identifying untapped potential and leveraging it to propel oneself or one's organization towards greater success. By doing so, individuals and organizations can multiply their achievements and leave a lasting impact on their industries and the world.

THE ULTIMATE COMPETITIVE WISDOM

EPILOGUE

Pursuing the Infinite Quest

As we approach the conclusion of "The Ultimate Competitive Wisdom", we find ourselves standing on the threshold of a new era. Throughout this journey, we have delved deep into the realms of strategy, innovation, and the unwavering pursuit of excellence. We have examined the competitive strategies of renowned competitors and well-known global companies, drawing lessons from their triumphs and defeats. Now, it is time to reflect upon the invaluable lessons we have learned and prepare ourselves for the infinite quest that lies ahead.

In our pursuit of competitive wisdom, we have come to realize that true success lies not only in achieving short-term victories but also in embracing a mindset that transcends momentary triumphs. The ultimate competitor understands that the quest for excellence is an ongoing journey—a continuous evolution of knowledge, skill, and adaptability.

Within the pages of this book, we have uncovered the power of strategic thinking, the importance of embracing change, and the value of resilience in the face of adversity. We have witnessed the transformative impact of innovation and the necessity of anticipating and embracing emerging trends.

But beyond the tactics and techniques lies a deeper understanding—a realization that true competitive wisdom is not limited to the realms of business and competition. It extends to every aspect of our lives, encouraging us to challenge ourselves, reach new heights, and continuously grow.

As we conclude this chapter of our collective quest, let us remember that competition is not merely about defeating others but

THE ULTIMATE COMPETITIVE WISDOM

also about elevating ourselves. It is about finding the strength within to push beyond our limits and discover our true potential.

The ultimate competitive wisdom lies in the synthesis of knowledge and intuition, analysis and creativity, courage and humility. It is a harmonious blend of strategy and adaptability, rooted in a deep understanding of oneself and the ever-changing landscape in which we operate.

As we bid farewell to these pages, let us carry forward the spirit of competition with grace, integrity, and empathy. Let us strive for excellence not only for personal gain but also for the betterment of our communities and the world at large.

Remember, the ultimate competitive wisdom does not reside in a finite destination but rather in the unending quest for improvement. It is a lifelong pursuit that requires constant introspection, learning, and growth.

May this book serve as a guide, igniting the fire of ambition and illuminating the path towards greatness. But ultimately, the true journey lies in your hands. Embrace the challenges that lie ahead, for it is through adversity that the seeds of greatness are sown.

Go forth, fellow seekers of wisdom, and let your competitive spirit soar. The infinite quest awaits.

REFERENCES

Ames, R. T. (1993). *Sun Tzu: The art of warfare*. Ballantine Books.

Barney, J. B. (1991). *Firm resources and sustained competitive advantage*. Journal of management, 17(1), 99-120.

Bass, B. M., & Riggio, R. E. (2006). Transformational Leadership (2nd ed.). Psychology Press.

Brown, T. (2009). *Change by Design: How Design Thinking Transforms Organizations and Inspires Innovation*. Harper Business.

Cleary, T. (2005). *The art of war: Sun Zi's military methods*. Shambhala Publications.

Cusumano, M. A. (2010). *Staying power: Six enduring principles for managing strategy and innovation in an uncertain world*. Oxford University Press.

Day, G. S., & Reibstein, D. J. (Eds.). (2004). *Wharton on dynamic competitive strategy*. John Wiley & Sons.

Denner, P. S. (2018). *Sun Tzu's the art of war and the wisdom of George S. Patton*. Naval War College Review, 71(3), 125-141.

Dierickx, I., & Cool, K. (1989). *Asset stock accumulation and sustainability of competitive advantage. Management science*, 35(12), 1504-1511.

Grant, R. M. (1996). *Toward a knowledge-based theory of the firm. Strategic management journal*, 17(S2), 109-122.

Garner, J. (2018). *Ancient Chinese military wisdom: Sun Tzu's lessons on strategy, opportunity, and risk*. Business Horizons, 61(5), 739-747.

Hamel, G., & Prahalad, C. K. (1994). *Competing for the future.* Harvard Business Review Press.

Hamel, G., & Prahalad, C. K. (2005). *Strategic intent.* Harvard Business Review, 83(7/8), 148-161.

Heller, M. (2001). *Reading Sun Tzu: The art of war and the philosophy of conflict.* Journal of Military Ethics, 1(1), 79-93.

Hill, C. W. L., & Jones, G. R. (2010). *Strategic management theory: An integrated approach.* Cengage Learning.

Hitt, M. A., Ireland, R. D., & Hoskisson, R. E. (2016). *Strategic management: Concepts and cases: Competitiveness and globalization.* Cengage Learning.

Kim, W. C., & Mauborgne, R. (2005). *Blue ocean strategy: From theory to practice.* California management review, 47(3), 105-121.

Kotler, P., & Keller, K. L. (2015). *Marketing management (15th ed.).* Pearson

Kouzes, J. M., & Posner, B. Z. (2017). The Leadership Challenge: How to Make Extraordinary Things Happen in Organizations (6th ed.). Wiley.

McNeilly, M. (2013). *Sun Tzu and the art of modern warfare.* Oxford University Press.

Mintzberg, H., Ahlstrand, B., & Lampel, J. (2008). *Strategy safari: A guided tour through the wilds of strategic management.* FT Press.

Porter, M. E. (1980). *Competitive strategy: Techniques for analyzing industries and competitors.* Free Press.

Porter, M. E. (1985). *Competitive advantage: Creating and sustaining superior performance.* Free Press.

Porter, M. E. (1996). *What is strategy?* Harvard Business Review, 74(6), 61-78.

Rumelt, R. P. (2011). *Good strategy/bad strategy: The difference and why it matters.* Crown Business.

Sawyer, R. D. (2015). *The seven military classics of ancient China.* Basic Books.

Sheppard, B. (2012). *The practical wisdom of Sun Tzu: The art of competitive decision making.* Journal of Business Strategy, 33(2), 4-13

Sun Tzu. (1910). *The art of war* (L. Giles, Trans.). Stackpole Books. (Original work published 5th century BC)

Sun Tzu. (2003). *The art of war.* Translated by Lionel Giles. Project Gutenberg.

Teece, D. J., Pisano, G., & Shuen, A. (1997). *Dynamic capabilities and strategic management.* Strategic management journal, 18(7), 509-533.

Treacy, M., & Wiersema, F. (1995). *The discipline of market leaders: Choose your customers, narrow your focus, dominate your market.* Addison-Wesley.

Tsai, S. (2015). *Adaptive analysis: Competing through the principles of Sun Tzu.* Strategy & Leadership, 43(5), 10-17

Verbeke, A. (2013). *International business strategy: Rethinking the foundations of global corporate success.* Cambridge University Press.

Washington, G. (1790). "To be prepared for war is one of the most effectual means of preserving peace." Quote from *the FIRST ANNUAL ADDRESS* TO BOTH HOUSES OF CONGRESS ON FRIDAY, JANUARY 08, 1790.

Wernerfelt, B. (1984). *A resource-based view of the firm.* Strategic management journal, 5(2), 171-180.

Wong, L. (2017). *The art of war by Sun Tzu: A modern interpretation for businesses.* World Scientific.

Yeh, K. S. (2017). *Sun Tzu's the art of war and strategic wisdom for senior executives.* Journal of Strategy and Management, 83-93.

Yim, H. M., & Yim, A. V. (2021). *WiseBook: A Practical Guide to Enlightenment Wisdom and Transformative Innovation Paradigm for Every Charismatic Leader.* The book is available on Amazon Kindle Book Store.

Yim, H. M., & Yim, A. V. (2022). *The Great Game XII: Eagle Vs. Dragon.* The book is available on Amazon Kindle Book Store.

Zhang, L. (2018). *A conceptual framework of Sun Tzu's art of war for strategic crisis management.* International Journal of Conflict Management, 29(1), 73-95.

Zhang, Y., & Huang, C. (2019). *An analysis of strategic thinking in Sun Tzu's the art of war and its implications for business.* Journal of Business Studies Quarterly, 11(3), 109-123.

Zook, C., & Allen, J. (2001). *Profit from the core: Growth strategy in an era of turbulence.* Harvard Business Press.

Zhu, K., & Wong, P. K. (2016). *Wisdom from ancient Chinese philosophy: The art of war and competitive advantage.* Business Horizons, 59(2), 225-234.

THE ULTIMATE COMPETITIVE WISDOM

INDEX

Achieving Exponential Success, 57

Adaptability and Agility, 7

Adaptability and Resilience, 2

Alliance With Empathy, 2

Art of Minimal Force, 54

Art of War, 1

Balancing Aggression and Protection, 51

Chaos Conceals Opportunity, 22

Collaboration and Cooperation, 3

Competitive Advantage, 2

Complacency Erodes Excellence, 41

Continuous Learning, 3

Design Thinking, 3

Embracing Competitive Wisdom, 1

Ethical Strategic Deception, 14

Greatest Competitive Edge, 29

Harmonizing Offense and Defense, 24

Informed Decision-Making, 2

Loyal Workforce Collaboration, 20

Maxim I, 5

Maxim II, 7

THE ULTIMATE COMPETITIVE WISDOM

Maxim III, 9

Maxim IV, 12

Maxim V, 14

Maxim VI, 17

Maxim VII, 20

Maxim VIII, 22

Maxim IX, 24

Maxim X, 27

Maxim XI, 29

Maxim XII, 31

Maxim XIII, 34

Maxim XIV, 37

Maxim XV, 41

Maxim XVI, 44

Maxim XVII, 48

Maxim XVIII, 51

Maxim XIX, 54

Maxim XX, 57

Maxim of Competitive Wisdom, 5, 7, 9, 12, 14, 17, 20, 22, 24, 27, 29, 31, 34, 37, 41, 44, 48, 51, 54, 57

Outmaneuvering Your Competition, 34

Peace Through Diplomacy, 2

Power Of Design Thinking Innovation, 27

Preparation And Decisiveness, 31

THE ULTIMATE COMPETITIVE WISDOM

Real-Life Scenario:

Airbnb: Revolutionizing Hospitality Through Design Thinking, 28

Amazon's Disruptive Success, 32

Apollo 11 Moon Landing: Conquering the Impossible, 55

Apple Vs. Microsoft: Success Through Strategic Deception, 15

Coca-Cola: A Model of Global Strategic Adaptability, 8

Dynamics Of Modern Warfare, 52

From Electric Cars to Space Exploration: Elon Musk's Journey of Seizing Opportunities, 58

Harnessing Collective Intelligence for Global Climate Action, 39

Netflix Vs. Blockbuster, 18

Nike Vs. Reebok: Strategic Maneuvers in The Sportswear Rivalry, 35

Operation Desert Storm: Triumph Through Strategy and Tactics, 45

Sacrifice And Success: The Inspiring Journey of Oprah Winfrey, 49

Samsung Vs. Apple, 10

Starbucks Family-like Culture, 4

Strategic Alliance of Apple And IBM, 13

Technology Industry Strategies: Lessons from Google and Amazon, 25

The Cuban Missile Crisis, 6

The Downfall of Kodak, 42

THE ULTIMATE COMPETITIVE WISDOM

The Michael Jordan Story, 30

The Story of 3M Post-It Notes, 23

Resource Optimization, 2

Sacrifice and Success, 48

Self-Awareness and Transformation, 9

Strategic Reconnaissance, 37

Strategic Unpredictability, 17

Strategy Development, 2

Synergy Of Strategy and Tactics, 44

Transformational Leadership, 3

ABOUT THE AUTHORS

Henry M. Yim is a talented young American writer, poet, math coach, inventor, strategist with a keen interest in Sun Tzu's Art of War. Henry is making waves in the world of STEM education and innovation. As co-founder of Henry STEM Academy, he provides free virtual STEM tutoring to children worldwide, covering subjects such as Geometry, Algebra I and II, and Calculus.

Henry's creativity extends to his inventions, including the "Jedimatics Fitness System ™" and "I-Pilot" design concept of the Autonomous Digital Vehicle (ADV). He is a keynote speaker and active volunteer in various programs, with his transformative innovation project, the "Cosmic Space Dome Greenhouse," even receiving the prestigious SpaceX Prize for its rigor in the scientific method with application to SpaceX and the aerospace field.

In addition, Henry is the co-author of the "WiseBook" on charismatic leadership, "Milky Way Poems" exploring the wonders of our galaxy, and "The Great Game XXI" promoting world peace. When he is not pursuing his many passions, Henry enjoys playing baseball, and his team won the Novi Youth Baseball League Championship. He resides in Michigan, USA.

THE ULTIMATE COMPETITIVE WISDOM

Alexandra V. Yim is an extraordinary young American inventor, writer, poet, and world peace advocate. She co-founded Henry STEM Academy and has a keen interest in science and world history. With a mission to promote freedom, justice, and peace globally, Alexandra recently launched the "World Peace Think Tank." She passionately urges every world leader to embrace the freedom of all members of the human family.

Alongside her academic achievements, Alexandra is recognized for her innovative project, the "Eco-friendly Greenhouse Ecosystem," designed to eliminate microplastics from oceans. She showcased this invention in the Michigan regional competition for the Invention Convention, coordinated by the Honors College of Oakland University. Alexandra has co-authored several noteworthy works, including the "WiseBook," which delves into charismatic leadership, "Milky Way Poems" that explore the wonders of our galaxy, and "The Great Game XXI," an advocate for world peace.

Alexandra generously shares her knowledge with other young students through her YouTube channel, "Jedimatics". On her channel, she not only teaches subjects like Geometry and Science but also crafts contemporary versions of moral stories.

Outside of her pursuits, Alexandra finds joy in playing softball, roller skating, and writing. She currently resides in Michigan, USA.

THE ULTIMATE COMPETITIVE WISDOM

Credits:

Cookie030307,CC-BY-SA 4.0 <https://creativecommons.org/licenses/by-sa/4.0>, via Wikimedia Commons (front cover)

THE ULTIMATE COMPETITIVE WISDOM

We would like to express our gratitude to all the readers who have purchased our new book, "The Ultimate Competitive Wisdom." Your support means a lot to us, and we hope the book has provided insightful and informative content on the topic of competitive strategy that drives success and winning.

Furthermore, we take this opportunity to encourage you to become advocates of world peace, particularly for the next generation of leaders. We firmly believe that each one of us can make a difference in creating a more peaceful world.

We eagerly await your views on the book and warmly welcome any feedback you may have. Please feel free to contact us at henryyim8@yahoo.com.

Additionally, we would like to inform our readers that a portion of the net proceeds from this book will be donated to non-profit organizations focused on supporting children's medical research, education, health, and humanitarian services worldwide.

Once again, we sincerely thank you for your support and wish you all the best in your journey towards a more peaceful world.

THE ULTIMATE COMPETITIVE WISDOM

THE ULTIMATE COMPETITIVE WISDOM

THE ULTIMATE COMPETITIVE WISDOM

www.ingramcontent.com/poc-product-compliance
Lightning Source LLC
Chambersburg PA
CBHW020454220526
45464CB00002B/986